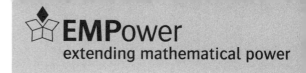

W9-CFO-111

Using Benchmarks

Fractions, Decimals, and Percents

STUDENT BOOK

TERC

Mary Jane Schmitt, Myriam Steinback,
Tricia Donovan, Martha Merson, and Marlene Kliman

Bothell, WA • Chicago, IL • Columbus, OH • New York, NY

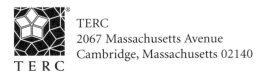

TERC
2067 Massachusetts Avenue
Cambridge, Massachusetts 02140

EMPower Research and Development Team
Principal Investigator: Myriam Steinback
Co-Principal Investigator: Mary Jane Schmitt
Research Associate: Martha Merson
Curriculum Developer: Tricia Donovan

Contributing Authors
Donna Curry
Marlene Kliman

Technical Team
Graphic Designer and Project Assistant: Juania Ashley
Production and Design Coordinator: Valerie Martin
Copyeditor: Jill Pellarin

Evaluation Team
Brett Consulting Group:
    Belle Brett
    Marilyn Matzko

EMPower™ was developed at TERC in Cambridge, Massachusetts. This material is based upon work supported by the National Science Foundation under award number ESI-9911410. Any opinions, findings, and conclusions or recommendations expressed in this publication are those of the authors and do not necessarily reflect the views of the National Science Foundation.

TERC is a not-for-profit education research and development organization dedicated to improving mathematics, science, and technology teaching and learning.

All other registered trademarks and trademarks in this book are the property of their respective holders.

http://empower.terc.edu

Printed in the United States of America
2 3 4 5 6 7 8 9 QDB 15 14 13 12 11

ISBN 978-0-07662-090-6
MHID 0-07-662090-5

Contents

Introduction

Welcome to EMPower

Students using the EMPower books often find that EMPower's approach to mathematics is different from the approach found in other math books. For some students, it is new to talk about mathematics and to work on math in pairs or groups. The math in the EMPower books will help you connect the math you use in everyday life to the math you learn in your courses.

We asked some students what they thought about EMPower's approach. We thought we would share some of their thoughts with you to help you know what to expect.

"It's more hands-on."

"More interesting."

"I use it in my life."

"We learn to work as a team."

"Our answers come from each other... [then] we work it out ourselves."

"Real-life examples like shopping and money are good."

"The lessons are interesting."

"I can help my children with their homework."

"It makes my brain work."

"Math is fun."

EMPower's goal is to make you think and to give you puzzles you will want to solve. Work hard. Work smart. Think deeply. Ask why.

Using This Book

This book is organized by lessons. Each lesson has the same format.

- The first page explains the lesson and states the purpose of the activity. Look for a question to keep in mind as you work.

- The activity page comes next. You will work on the activities in class, sometimes with a partner or in a group.

- Look for shaded boxes with additional information and ideas to help you get started if you become stuck.

- Practice pages follow the activities. These practices will make sense to you after you have done the activity. The three types of practice pages are

  Practice: Provides another chance to see the math from the activity and to use new skills.

  Extension: Presents a challenge with a more difficult problem or a new but related math idea.

  Test Practice: Asks a number of multiple-choice questions and one open-ended question.

In the *Appendices* at the end of the book, there is space for you to keep track of what you have learned and to record your thoughts about how you can use the information.

- Use notes, definitions, and drawings to help you remember new words in *Vocabulary*, pages 101–102.

- Answer the *Reflections* questions after each lesson, pages 103–108.

Tips for Success

Where do I begin?

Many people do not know where to begin when they look at their math assignments. If this happens to you, first try to organize your information. Read the problem. Start a drawing to show the situation.

Much of this unit is about parts and wholes.

Ask yourself:

> *What makes up the whole group? What number is just part of the group?*

Another part of getting organized is figuring out what skills are required.

Ask yourself:

> *What do I already know? What do I need to find out?*

Write down what you already know.

I cannot do it. It seems too hard.

Make the numbers smaller or friendlier. Try to solve the same problem with the benchmark fraction $\frac{1}{2}$.

Ask yourself:

> *Have I ever seen something like this before? What did I do then?*

You can always look back at another lesson for ideas.

Am I done?

Don't walk away yet. Check your answers to make sure they make sense.

Ask yourself:

Did I answer the question?

Does the answer seem reasonable? Do the conclusions I am drawing seem logical?

Check your math with a calculator. Ask others whether your work makes sense to them.

Practice

Complete as many of the practice pages as you can to sharpen your skills.

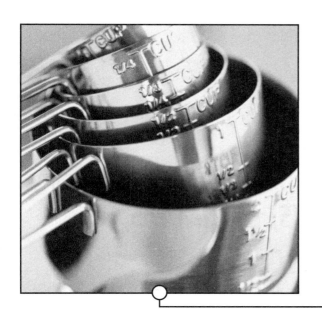

Opening the Unit: Using Benchmarks

> ## *What are the most common fractions?*

In this session, you are introduced to the unit *Using Benchmarks: Fractions, Decimals, and Percents.* You also explore the fraction $\frac{1}{2}$ (**one-half**) and share what you know about the other common fractions.

Common fractions include $\frac{1}{2}$, $\frac{1}{4}$, $\frac{3}{4}$, and $\frac{1}{10}$. This unit gets you thinking about these fractions in many different ways so they become as familiar to you as the numbers you use to count.

Activity 1: Making a Mind Map

Make a Mind Map using words, numbers, pictures, or ideas that come to mind when you think about *fractions*, *decimals*, and *percents*.

Activity 2: I Will Show You $\frac{1}{2}$!

Part 1

Use something in the classroom—or a group of things—to demonstrate one-half. Show or draw $\frac{1}{2}$ in at least two ways.

Object(s) _____

What is the *whole?*
What is the *part?*

Object(s) _____

What is the *whole?*
What is the *part?*

Activity 2: I Will Show You $\frac{1}{2}$! *(continued)*

Part 2

1. **a.** Fill in the blanks.

Whole	Part	Fraction	Equivalent
1 day 24 hours	12 hours	$\frac{12}{24}$	$\frac{1}{2}$
2 days 48 hours	24 hours	$\frac{24}{48}$	
5 days 120 hours	60 hours		$\frac{1}{2}$
10 days 240 hours	120 hours		

    b. What is the pattern between the part numbers and the whole numbers in this chart?

2. Think about a dollar, made up of 100 pennies, and complete the table below.

Number	Whole	Part	Fraction	Equivalent
$ 1	100 pennies	50 pennies	$\frac{50}{100}$	$\frac{1}{2}$
$ 2				
$ 5				
$10				

3. Write a rule for finding $\frac{1}{2}$ of any number.

Activity 3: Initial Assessment

Your teacher will show you some problems and ask you to check off how you feel about your ability to solve each problem.

____ Can do   ____ Don't know how   ____ Not sure

Have you ever noticed that every new place where you work has its own words or specialized vocabulary? This is true of topics in math too. In every lesson you will be introduced to some specialized vocabulary. Do not worry if you see words in the problems that you do not recognize. You can write some words down and look them up later or learn as you go.

Using Benchmarks: Fractions, Decimals, and Percents Unit Goals

What are your goals regarding the study of fractions, decimals, and percents? Review the following goals in this book. Then think about your own goals and record them in the space provided.

- Describe part-whole situations in terms of fractions.

- Use objects, diagrams, number-line segments, and arrays to represent part-whole situations.

- Determine if a wide variety of fractions are more than, less than, or equal to the benchmark fractions $\frac{1}{2}$, $\frac{1}{4}$, $\frac{3}{4}$, and $\frac{1}{10}$.

- Read benchmark decimals such as 0.1 as fractions.

- Connect percent names with benchmark fractions and decimals.

My Own Goals

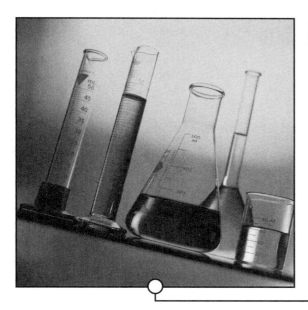

More Than, Less Than, or Equal to One-Half?

Where's the halfway mark?

In our everyday lives, we often hear mentioned or use ourselves the fraction $\frac{1}{2}$. Half is called a "**benchmark**" fraction because it is used so often as a basis for comparing amounts. In this lesson, you will find fractions of various amounts and then figure out if they are more than, less than, or equal to one-half.

Activity 1: Stations—Comparing Fractions to $\frac{1}{2}$

At each station, identify

- The part and
- The whole.

Write the fraction represented at the station, and then answer these questions:

- Does the fraction equal one-half?
- Is it greater than one-half?
- Is it less than one-half?

Think about how you know the answers to the questions.

Complete the following table:

Comparing Fractions to $\frac{1}{2}$

Item at Each Station	Part	Whole	Fraction	Equal to (=) $\frac{1}{2}$ Greater Than (>) $\frac{1}{2}$ Less Than (<) $\frac{1}{2}$
Station 1: one month	16 days			
Station 2:				
Station 3:				
Station 4:				
Station 5:				

Activity 2: Is It Half?

In the problems below, you will practice thinking about and writing fractions that are more than, less than, or equal to one-half.

1. Sherri dropped a box of 80 crackers. She took them all out and counted 60 broken crackers.

Fraction for the Total	Fraction Described in the Story	Fraction for Half of the Total

The number of broken crackers was

    a. More than $\frac{1}{2}$ the whole box.

    b. Less than $\frac{1}{2}$ the whole box.

    c. $\frac{1}{2}$ the whole box.

Show your reasoning with a sketch or number line.

2. Vernon had $15. He gave $7 to his sister.

Fraction for the Total	Fraction Described in the Story	Fraction for Half of the Total

Vernon gave away

    a. More than 50% of his money.

    b. Less than 50% of his money.

    c. 50% of his money.

Show your reasoning with a sketch or number line.

3. A ream of paper has 500 sheets. Allie used up 228 sheets.

Fraction for the Total	Fraction Described in the Story	Fraction for Half of the Total

Allie used up

   a. More than 50% of the ream.

   b. Less than 50% of the ream.

   c. 50% of the ream.

Show your reasoning with a sketch or number line.

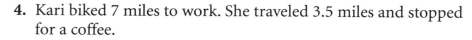

4. Kari biked 7 miles to work. She traveled 3.5 miles and stopped for a coffee.

Fraction for the Total	Fraction Described in the Story	Fraction for Half of the Total

Kari was

   a. More than halfway to work.

   b. Less than halfway to work.

   c. Halfway to work.

Show your reasoning with a sketch or number line.

5. Out of the dozen muffins we started with, only three are left.

Fraction for the Total	Fraction Described in the Story	Fraction for Half of the Total

We now have

a. More than $\frac{1}{2}$ of the muffins.

b. Less than $\frac{1}{2}$ of the muffins.

c. $\frac{1}{2}$ of the muffins.

Show your reasoning with a sketch or number line.

6. Nanda was making bread. The recipe called for 3 cups of flour. She measured out $1\frac{1}{2}$ cups of flour.

Fraction for the Total	Fraction Described in the Story	Fraction for Half of the Total

Nanda measured out

a. More than $\frac{1}{2}$ the flour she needs.

b. Less than $\frac{1}{2}$ the flour she needs.

c. $\frac{1}{2}$ the flour she needs.

Show your reasoning with a sketch or number line.

Practice: Half the Size

Show the amount indicated in each of the following problems.

1. Shade $\frac{1}{2}$ of 8 ounces (oz.).

2. Create a flavor that is half as popular as vanilla. Show it on the graph.

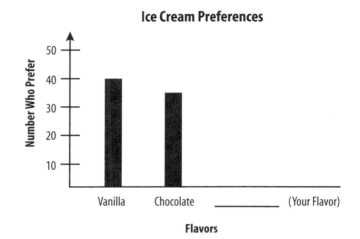

3. Show half the candles in two different ways.

a.

b.

4. a. Rinaldo is a soccer fan. He roots for Brazil and wants to paint the wall in his bedroom half yellow and half green. Use the following picture to show him how to do that. Note the rectangle where his window is located.

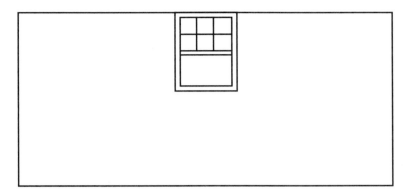

b. How do you know that you have figured out a way to paint the room half yellow and half green?

c. Could you have shown it another way? How?

5. Traci took a trip last week. She traveled in a circle, starting and ending at home.

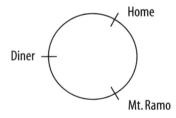

a. She said the whole trip took her six and a half hours. About where do you think she was at the halfway mark? How do you know?

b. Traci drove a total of 250 miles. At the halfway mark, how many miles had she driven? How do you know?

Practice: Why Is 50% a Half?

How would you explain to a friend that 50% equals $\frac{1}{2}$? Use pictures, words, or the grid below to support your explanation.

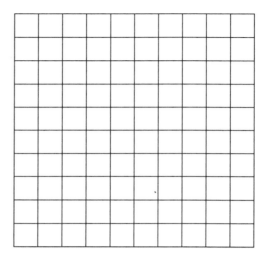

Practice: Find Half of It

Calculate half of each amount. Some problems are easier to do than others. The first one is done for you.

1. 4 million   <u> 2 million </u>

2. 750,000 people in Boston   <u>   </u>

3. 88-year-old aunt   <u>   </u>

4. 16-oz. chocolate bar (one pound)   <u>   </u>

5. 25-percent increase   <u>   </u>

6. 30-day billing cycle   <u>   </u>

7. 90° angle   <u>   </u>

8. 400 years of oppression   <u>   </u>

9. 1 mile of swimming, or 72 laps   <u>   </u>

10. 54 days until summer vacation   <u>   </u>

Practice: Choose an Amount

1. An amount that is more than $\frac{1}{2}$ a 16-oz. chocolate bar is _____.

2. An amount that represents more than $\frac{1}{2}$ the days in a 30-day billing cycle is _____.

3. An amount for an angle that is less than $\frac{1}{2}$ a 90° angle is _____.

4. An amount that is more than $\frac{1}{2}$ of 88 years is _____.

5. An amount greater than 50% of 4 million is _____.

6. An amount greater than 50% of 750,000 people in Boston is _____.

7. An amount that is less than $\frac{1}{2}$ of 3 cups of water is _____.

8. In general, to find $\frac{1}{2}$ (or 50%) of any amount, this is what I do:

9. If I know what half a total amount equals, and I want to find the total, this is what I do:

Practice: More "Is It $\frac{1}{2}$?" Problems

1. Irina had a 21-inch ribbon. She cut off 10 inches for her daughter. She then had

   a. More than $\frac{1}{2}$ the ribbon left.

   b. Less than $\frac{1}{2}$ the ribbon left.

   c. $\frac{1}{2}$ the ribbon left.

   Show how you figured out your answer.

2. The Marquez-Brown family owns a ranch with 1,300 acres of land. They plan to sell 600 acres. They will sell

   a. More than $\frac{1}{2}$ their land.

   b. Less than $\frac{1}{2}$ their land.

   c. $\frac{1}{2}$ their land.

   Show how you figured out your answer.

3. Lourdes was traveling at 55 mph (miles per hour) when traffic suddenly slowed to 25 mph. Lourdes' traveling speed became

   a. More than $\frac{1}{2}$ of what it had been.

   b. Less than $\frac{1}{2}$ of what it had been.

   c. $\frac{1}{2}$ of what it had been.

   Show how you figured out your answer.

Practice: What Is the Whole?

You have looked at whole amounts to figure out what one-half equals. In the following problems you are given half an amount and must find the whole.

> Remember, there are several ways to indicate one half:
> $\frac{1}{2}$ 50%   0.5

1. Complete the table below:

$\frac{1}{2}$ the Number	The Whole Number	Fraction for the Whole
15		
75		
$3\frac{1}{2}$		
335		
22.5		

2. Frank spent half his paycheck on an overdue phone bill. He paid $160. How much was his whole paycheck? Show your work.

3. Mary Jane ate half her recommended daily allowance of calories at lunch when she consumed 750 calories. How many calories per day are there in Mary Jane's recommended daily allowance? Show your work.

4. Juan's two-year-old son weighs half the amount his four-year-old son weighs. His two-year-old weighs 19.5 lbs. How much does his four-year-old son weigh? Show your work.

5. Fifty percent of Melda's classmates are immigrants. Melda says there are 18 immigrants in her class. How many of her classmates are *not* immigrants? How many people altogether are there in Melda's class? Show your work.

Practice: Which Is Larger?

Use what you know about finding half of a whole to compare the following fractions.

> Comparing fractions is easier when you think in terms of benchmarks.
> Is the fraction less than $\frac{1}{2}$? Is the fraction greater than $\frac{1}{2}$?

In the fraction pairs below, circle the larger fraction. Explain how you know which fraction is larger.

Note: One pair of fractions is equal. How did you pick out that pair?

Fraction Pairs	**How I know which is larger....**
1. $\frac{2}{5}$ or $\frac{5}{8}$	
2. $\frac{3}{4}$ or $\frac{1}{3}$	
3. $\frac{8}{16}$ or $\frac{6}{12}$	
4. $\frac{4}{7}$ or $\frac{4}{9}$	

Extension: Half a Million?

On Sunday, *The Daily Star* newspaper is distributed in three towns: Crystal, Jackson, and Santa Linda.

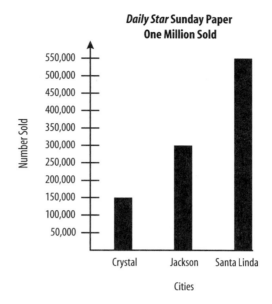

1. How many copies of *The Daily Star* were sold on Sunday?

   _____

2. The number of papers sold in Santa Linda was

   a. More than $\frac{1}{2}$ a million.

   b. Less than $\frac{1}{2}$ a million.

   c. $\frac{1}{2}$ a million.

3. The number of papers sold in Crystal and Jackson together was

   a. More than $\frac{1}{2}$ a million.

   b. Less than $\frac{1}{2}$ a million.

   c. $\frac{1}{2}$ a million.

4. The number of papers sold in Crystal was

   a. More than $\frac{1}{2}$ the number sold in Jackson.

   b. Less than $\frac{1}{2}$ the number sold in Jackson.

   c. $\frac{1}{2}$ the number sold in Jackson.

5. Which two towns together sold more than half a million papers?

   How do you know?

6. Write a sentence about the graph data that uses the term "50%."

1. Five people shared two pizzas so that everyone received the same amount. Each person's share was

   (1) $\frac{1}{2}$ a pizza.

   (2) More than half a pizza.

   (3) Less than half a pizza.

   (4) 0.5 of a pizza.

   (5) More than 50% of a pizza.

2. There are more than 600 species of bacteria in a human's mouth. If you destroyed 275 of those species, you would have eliminated

   (1) $\frac{1}{2}$ the bacteria species.

   (2) More than $\frac{1}{2}$ the bacteria species.

   (3) Less than $\frac{1}{2}$ the bacteria species.

   (4) All the bacteria species.

   (5) 50% of the bacteria species.

3. Penny said that one year the average Dow Jones stock-market posting showed a gain, going from about 10,000 points to 12,000 points. According to Penny, that year the stock market rose

   (1) More than $\frac{1}{2}$.

   (2) Less than $\frac{1}{2}$.

   (3) 50%.

   (4) $\frac{1}{2}$.

   (5) More than 50%.

4. A newspaper reported that "Today, nine out of 10 American homes are equipped with a (smoke) detector, and residential fire deaths have been cut in half since 1973. 'That's 50,000 people that didn't die'" (*Boston Globe*, June 6, 2004, p. E2). According to the paper, how many people did die from fire-related injuries during that time?

   (1) 150,000

   (2) 100,000

   (3) 50,000

   (4) 25,000

   (5) 10,000

5. A study found that people who drink at least one cup of tea a day are half as likely to develop ulcers as those who do not. Which, if any, of the following conclusions can be made based solely on this information?

   (1) Coffee causes ulcers.

   (2) Drinking tea prevents all ulcers.

   (3) Tea drinkers develop ulcers 50% more often than non-tea drinkers.

   (4) Tea drinkers develop ulcers nearly 50% more often than non-tea drinkers.

   (5) Tea drinkers develop ulcers nearly 50% less often than non-tea drinkers.

6. Three partners own a small business firm. Tamika owns 10% of the company, Eduardo owns 40%, and Renee owns the remainder. Last year the company made $200,000 in profits. What fraction of the company does Renee own?

Half of a Half

> *How do you recognize a quarter?*

Sometimes it is not enough to know what half of something is; the amount you need to find is **one-fourth**, or a half of a half.

The fraction $\frac{1}{4}$ is a signal; what action will you take when you see $\frac{1}{4}$? In this lesson you will build on what you know about one-half to show fourths in different ways.

Activity 1: $\frac{1}{4}$ Wasted

You can refer to one-fourth in several ways: 1/4, 0.25, 25%, or one-quarter. If you waste, lose, or use 1/4 of something, you have 3/4 left.

$$\frac{1}{4} \text{ wasted} \quad + \quad \frac{3}{4} \text{ left} \quad = \quad \frac{4}{4}, \text{ the whole}$$

- In Problems 1 and 2, you are given a total and asked to find $\frac{1}{4}$ of it that was wasted and the 3/4 left.

- In Problems 3, 4, and 5, you are given the amount equal to $\frac{1}{4}$ of the total and asked to find either the whole amount or the amount left over, 3/4 of the total.

- Use a picture or number line to illustrate your thinking for at least Problems 1a*, 2c*, 3b*, and 4a*.

1. Mad-Cow Meat Scare

   a.* 20 pounds of meat total

   $\frac{1}{4}$ wasted = _____ $\frac{3}{4}$ left = _____

   b. 82 pounds of meat total

   $\frac{1}{4}$ wasted = _____ $\frac{3}{4}$ left = _____

   c. 2,000 pounds of meat total

   $\frac{1}{4}$ wasted = _____ $\frac{3}{4}$ left = _____

2. Easy Come, Easy Go Money

   a. $10.00 in all

   $\frac{1}{4}$ wasted = _____ $\frac{3}{4}$ left = _____

   b. $5.00 in all

   $\frac{1}{4}$ wasted = _____ $\frac{3}{4}$ left = _____

   c.* $90.00 paycheck total

   $\frac{1}{4}$ wasted = _____ $\frac{3}{4}$ left = _____

3. Time Is Money

   a. One-quarter of my available work time is wasted in commuting an hour. What is my total available work time?

   b.*One-quarter of my available work time is wasted in commuting two and a half (2.5) hours. What is my total available work time?

   c. One-quarter of my available work time is wasted in commuting six and a half (6.5) hours. What is my total available work time?

   d. One-quarter of my available work time is wasted in commuting 48 hours. What is my total available work time?

4. Flooded Basement

   a.* $\frac{1}{4}$ of the books ruined = 24 books. How many books were there to start?

   b. 25% of the books ruined = 120 books. How many books were there to start?

   c. $\frac{1}{4}$ of the books ruined = 57 books. How many books were there to start?

5. Movie Set Auditions

   a. One hundred and twenty people were called to audition. Twenty-five percent (25%) of the actors were hired; 25% (or $\frac{1}{4}$) = _____. Seventy-five percent (75%) of the actors were not hired; 75% (or $\frac{3}{4}$) = _____.

   b. Twenty-five percent (25%) of the actors was a total of nine people. How many were called in all? _____

   c. Twenty-five percent (25%), or $\frac{1}{4}$, of those called but *not* hired = 25 people. How many were called in all? _____ How many were hired to work on the set? _____

Activity 2: Is It Really a Quarter?

- Read the following headline information. What numbers are missing?

- The reports do not give total or one-quarter amounts. Make up some numbers as examples that could stand for totals or quarters in a full report.

- Check your work: Is the amount really a quarter?

- Demonstrate your reasoning using pictures, words, or numbers.

- Write a brief paragraph that might follow your headline. Include your diagram or number line.

1.

March 31, 2004

Attack Readiness Drops

A recent poll finds fewer Americans are following the government's advice. One-quarter of those polled have a designated "safe room."

Source: *USA TODAY*, March 2004

2.

Household Break-Ins Drop By More Than One-Quarter

3.

Extra Calorie Price Alert

Researchers report that at Dandy Donuts bakeries, the price for the classic size donut jumped about 25%, but the number of calories jumped more than 100%.

Practice: What Makes It a Quarter?

Look at the grid below. Explain how you know that the shaded portion is a quarter, $\frac{1}{4}$, 0.25, or 25%.

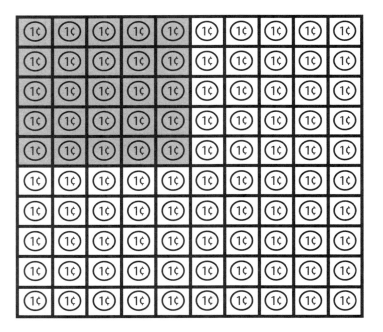

Practice: Show Me $\frac{1}{4}$!

Shade the portion that equals $\frac{1}{4}$ of each of the following shapes or sets of objects, and fill in the blanks.

1.

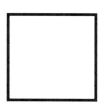

    a. The total number of pieces (the whole) is _____.

    b. The number of shaded pieces (the part) is _____.

    c. The fraction is _____.

2.

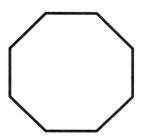

    a. The total number of pieces (the whole) is _____.

    b. The number of shaded pieces (the part) is _____.

    c. The fraction is _____.

3.

    a. The total number of pieces (the whole) is _____.

    b. The number of shaded pieces (the part) is _____.

    c. The fraction is _____.

Mark $\frac{1}{4}$ of the distance from zero in each numberline segment below.

4.

a. The total number of units in the number line (the whole) is _____.

b. The number of shaded units (the part) is _____.

c. The fraction is _____.

5.

a. The total number of units in the number line (the whole) is _____.

b. The number of shaded units (the part) is _____.

c. The fraction is _____.

6.

a. The total number of units in the number line (the whole) is _____.

b. The number of shaded units (the part) is _____.

c. The fraction is _____.

Practice: $\frac{1}{4}$ Measurements

1. Mark the line that shows $\frac{1}{4}$ of 1 cup.

   $\frac{1}{4}$ of 1 c. = _____

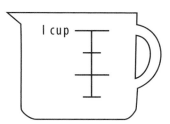

2. Mark the line that shows $\frac{1}{4}$ of 4 cups.

   $\frac{1}{4}$ of 4 c. = _____

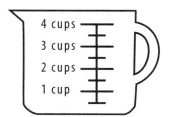

3. Mark the line that shows $\frac{1}{4}$ of 2 cups.

   $\frac{1}{4}$ of 2 c. = _____

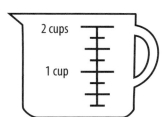

4. Mark the line that shows $\frac{1}{4}$ of 12 cm.

   $\frac{1}{4}$ of 12 cm = _____ cm

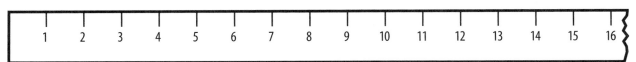

5. Mark the line that shows $\frac{1}{4}$ of 16 cm.

   $\frac{1}{4}$ of 16 cm = _____ cm

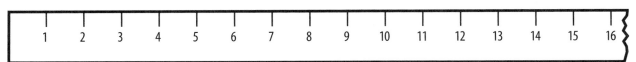

6. $\frac{1}{4}$ of 5 pounds = _____

7. $\frac{1}{4}$ of 10 kilograms = _____

8. $\frac{1}{4}$ of a day = _____ hours

9. Mark the line that shows 1/4 of an hour.

$\frac{1}{4}$ of 60 minutes = _____ minutes

10. Use an example of your own to explain how to find $\frac{1}{4}$ (25%) of a number.

Practice: How Many, How Far?

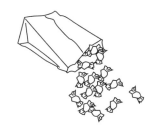

1. Chen and Lucia agreed to share a bag of candies. Chen took 45 candies and Lucia took 15. Kurt was watching and said, "This sharing is not fair! That is not a 50-50 split!"

   a. How many candies were in the bag? How do you know?

   b. What fraction of the candies did Lucia get? (Label the part and the whole.)

   c. Is Kurt right? Explain.

   d. Suggest a different way of sharing this bag of candies.

   e. Would Kurt say that this way of sharing is fair? Why?

2. On Saturdays Enrique walks 18 blocks to the Farmers' Market. He takes his granddaughter Becky with him. She asks, "Are we there yet? Are we there yet?" even though they have just started.

He says, "I will tell you when we are one-fourth the way there, then halfway there, and you will guess when we have only one block to go."

   a. At what block will Enrique tell Becky that they are one-fourth the way there?

   b. At what block will Enrique tell Becky that they are halfway there?

   c. How do you know? Use a diagram to show your reasoning.

3. Ginger wants to bake brownies. Her recipe calls for eight eggs. She looks in her refrigerator and notices that she only has two eggs. She decides to bake a smaller amount and reduces the recipe.

a. What fraction of the recipe is she making? How do you know?

b. Show how the amounts in her recipe change:

Super Gooey Brownies	Brownies with Only 2 Eggs
1 c. (cup) butter	
8 eggs	2 eggs
3 c. sugar	
2 c. flour	
2 tsp. vanilla	
6 oz. unsweetened chocolate	

c. Which ingredients were easy to adjust? Why?

d. Which ingredients were hard to adjust? Show how you solved these problems.

4. In June 2004, it was reported that immigrants were filling three out of 10 new jobs in the United States. Was this more than, less than, or exactly $\frac{1}{4}$ of the new jobs? Show your method of reasoning.

5. One-fourth of Shana's hourly wages go to cover healthcare benefits. Shana makes $12 an hour.

   a. How many dollars an hour does Shana contribute to healthcare benefits? Show your method of reasoning.

   b. If she gets a raise and the cost of health insurance stays the same, will Shana pay more or less than one-fourth of her new wage? Show your method of reasoning.

6. A special education fund is fully funded at $200 million. If legislators want to cut 25%, or $\frac{1}{4}$, of that budget, how much will they cut? Show your method of reasoning.

Practice: Comparing Fractions to $\frac{1}{4}$

1. For each example in the table below, identify the *part* and the *whole*. Then write the fraction represented by the example and answer the following questions:

   - Is it less than one-fourth?

   - Does the fraction equal one-fourth?

   - Is it greater than one-fourth?

Comparing Fractions to $\frac{1}{4}$

Example	Part	Whole	Fraction	Equal to (=) $\frac{1}{4}$ Greater Than (>) $\frac{1}{4}$ Less Than (<) $\frac{1}{4}$
a. 8 days of rain out of 30	_____ days	_____ days		
b. 125 yards run in the 440-yd. dash				
c. 20 minutes out of an hour-long test.				
d. 1,320 feet (′) walked out of a mile (5,280′)				

2. Choose one example and explain with words, diagrams, and/or a number line how you determined whether the fraction was less than, more than, or equal to $\frac{1}{4}$.

Extension: Which Is Larger?

Use what you know about finding one-fourth and one-half of a whole to compare the following fractions.

> Imagine that the fractions explain how many people attended two different classes.
>
> Circle the fraction that shows which class had better attendance.

Circle the larger fraction in the fraction pairs below. Explain how you know which is larger.

Fraction Pairs

1. $\frac{4}{9}$ or $\frac{2}{8}$

2. $\frac{3}{10}$ or $\frac{5}{8}$

3. $\frac{3}{12}$ or $\frac{3}{16}$

4. $\frac{1}{3}$ or $\frac{3}{8}$

How I know which is larger....

Test Practice

1. A glass-recycling bin holds 200 bottles. There are 50 green bottles in the bin. What fraction of the bottles is green?

   (1) $\frac{1}{200}$

   (2) $\frac{1}{4}$

   (3) $\frac{1}{2}$

   (4) $\frac{3}{4}$

   (5) $\frac{50}{50}$

2. There are 150 clear glass bottles in a recycling bin. These are $\frac{1}{4}$ of the total bottles in the bin. How many bottles are in the bin?

   (1) 150

   (2) 154

   (3) 300

   (4) 450

   (5) 600

3. Guy plans to raise $1,000 for the asthma fund. To figure out the amount that represents one-fourth of his goal, Guy should

   (1) Divide $1,000 by 4.

   (2) Find $\frac{1}{2}$ of 1,000 and multiply by 2.

   (3) Multiply 1,000 by 4.

   (4) Divide 1,000 by 2.

   (5) Divide $250 by 2.

4. In the graph below, which age group made up about $\frac{1}{4}$ of the total voters?

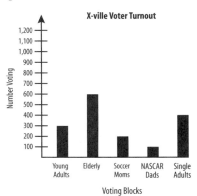

   (1) Soccer moms

   (2) Young adults

   (3) Elderly

   (4) NASCAR dads

   (5) Single adults

5. Ina installs $\frac{1}{4}$ of the fencing for the following dog pen and Sheila installs the rest. How many feet of fencing does Ina install?

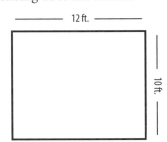

   (1) 10

   (2) 11

   (3) 12

   (4) 22

   (5) 33

6. Out of the 100 United States Senators, 25% were absent for a recent roll call vote. How many were present?

Three Out of Four

> *What does three-fourths mean?*

When you think about $\frac{3}{4}$ of any amount, you can tell two things right away: The amount is more than half ($\frac{2}{4}$) and less than the whole ($\frac{4}{4}$). In this lesson, you will use different ways to find **three-fourths** of any amount.

Activity 1: Seats for $\frac{3}{4}$

Architects designed the cafeteria for each building in the Plaza Complex to hold $\frac{3}{4}$ of the building's workers. How many people can sit in each cafeteria?

Building A	48 workers
Building B	100 workers
Building C	240 workers

1. How many people will the cafeteria in your building seat? (Use the building your teacher assigned to you.)

2. What fraction represents all the workers in your building?

3. What fraction represents the workers who can sit in the cafeteria of your building at one time? _____

   How do you know this equals 75% of the workers?

4. What fraction represents the workers who *cannot* sit in the cafeteria of your building when it is full?

5. What steps did you use to find three-fourths?

6. Make a number line to show your reasoning.

7. Show your reasoning on an array where each block represents one worker.

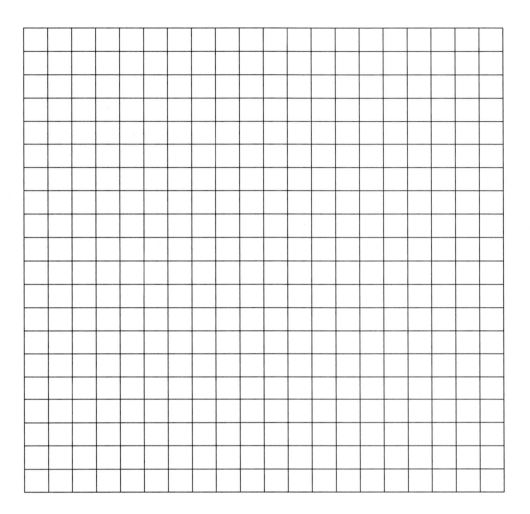

Activity 2: Where Are You From?

One day people in the office cafeteria were talking about which parts of the world they came from. They took a poll and found that

- Four people were from Africa;
- Seven people were from Asia;
- Thirteen people were from Central America;
- Eight people were from the United States.

1. Use the fractions $\frac{1}{4}$, $\frac{1}{2}$, and $\frac{3}{4}$ to describe the cafeteria groups by writing two true statements and one false statement. Remember to consider both the parts and the whole. You can also use the phrases "more than" and "less than."

   a. True:

   _____

   _____

   _____

   _____

   _____

   b. True:

   _____

   _____

   _____

   _____

   _____

   c. False:

   _____

   _____

   _____

   _____

   _____

2. **a.** The number of people from Central America *and* Asia is

   (1) Less than $\frac{3}{4}$ of the group.

   (2) More than $\frac{3}{4}$ of the group.

   (3) $\frac{3}{4}$ of the group.

   b. How do you know?

3. **a.** The number of people from Central America, Africa, *and* Asia is

   (1) Less than $\frac{3}{4}$ of the group.

   (2) More than $\frac{3}{4}$ of the group.

   (3) $\frac{3}{4}$ of the group.

   b. How do you know?

Practice: Show Me $\frac{3}{4}$!

Shade the portion that equals $\frac{3}{4}$ of each of the following shapes or sets of objects, and fill in the blanks.

1.

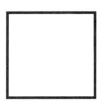

    a. The total number of pieces (the whole) is _____.

    b. The number of pieces shaded (the part) is _____.

    c. The fraction is _____.

2.

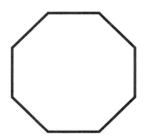

    a. The total number of pieces (the whole) is _____.

    b. The number of pieces shaded (the part) is _____.

    c. The fraction is _____.

3.

    a. The total number of parts (the whole) is _____.

    b. The number of parts shaded (the part) is _____.

    c. The fraction is _____.

Mark $\frac{3}{4}$ of the distance from zero in each number line segment below.

4.

|———|———|———|———|
0   4

    a. The total number of units in the number line (the whole) is _____.

    b. The number of shaded units (the part) is _____.

    c. The fraction is _____.

5.

|———|———|———|———|———|———|
0   1   2   3   4   5   6

    a. The total number of units in the number line (the whole) is _____.

    b. The number of shaded units (the part) is _____.

    c. The fraction is _____.

6.

|—|—|—|—|—|—|—|—|—|—|—|—|—|—|—|—|
0 1 2 3 4 5 6 7 8 9 10 11 12 13 14 15 16

    a. The total number of units in the number line (the whole) is _____.

    b. The number of shaded units (the part) is _____.

    c. The fraction is _____.

Practice: $\frac{3}{4}$ Measurements

In each problem below, shade or circle $\frac{3}{4}$ of the total.

1. Mark the line that shows $\frac{3}{4}$ of 1 cup.

   $\frac{3}{4}$ of 1 c. equals _____

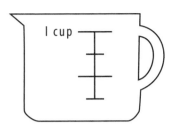

2. Mark the line that shows $\frac{3}{4}$ of 4 cups.

   $\frac{3}{4}$ of 4 c. = _____

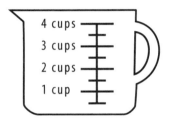

3. Mark the line that shows $\frac{3}{4}$ of 2 cups.

   $\frac{3}{4}$ of 2 c. = _____

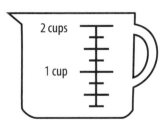

4. Mark the line that shows $\frac{3}{4}$ of 12 cm.

   $\frac{3}{4}$ of 12 cm = _____ cm

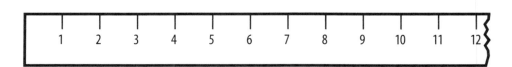

5. Mark the line that shows $\frac{3}{4}$ of 16 cm.

$\frac{3}{4}$ of 16 cm = _____ cm

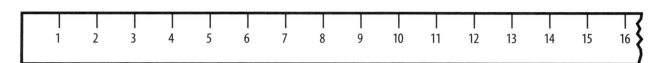

6. $\frac{3}{4}$ of 6 pounds = _____

7. $\frac{3}{4}$ of 12 kilograms = _____

8. $\frac{3}{4}$ of a day = _____ hours

9. Mark the line that shows $\frac{3}{4}$ of an hour.

$\frac{3}{4}$ of 60 minutes = _____ minutes

10. Use an example of your own to explain how to find $\frac{3}{4}$ (75%) of a number.

Practice: Where to Place It?

For each problem, first mark the given fraction on the line. Then circle the correct answer for whether the fraction is less than (<), equal to (=), or greater than (>) $\frac{3}{4}$.

1. $\frac{4}{14}$ is less than (<)   equal to (=)   greater than (>) $\frac{3}{4}$.

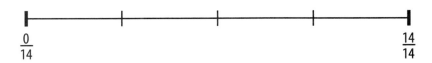

2. $\frac{15}{20}$ is less than (<)   equal to (=)   greater than (>) $\frac{3}{4}$.

3. $\frac{2}{3}$ is less than (<)   equal to (=)   greater than (>) $\frac{3}{4}$.

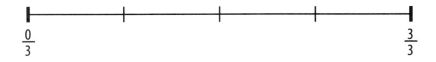

4. $\frac{45}{60}$ is less than (<)   equal to (=)   greater than (>) $\frac{3}{4}$.

5. Choose one of the problems above and describe how you compared the fractions to reach your conclusions.

Practice: More "How Many, How Far?" Problems

1. Valerie and Sol plan to raise money for the local hospital by getting pledges for the Hospital Walk-a-Thon. Valerie is responsible for walking $\frac{1}{4}$ the distance and raising $\frac{1}{4}$ of the pledges. Sol is responsible for walking $\frac{3}{4}$ the distance and raising $\frac{3}{4}$ of the pledges.

   a. The Hospital Walk-a-Thon covers 24 miles. Valerie walks 5 miles and stops. Did she walk as far as she planned? Why or why not?

      Use a diagram or number line to support your reasoning.

   b. Together Valerie and Sol raised $3,000. On the graph below, show how much money each person should have raised as his or her share of the total. How did you determine your answer?

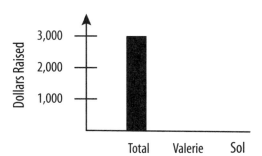

2. Joe and his Aunt Ellie walk to the movies every Thursday night. It is 20 blocks from their apartment building to the movie theater. They always stop $\frac{3}{4}$ of the way there to buy snacks.

   a. At what block do they stop? _____

   b. If the total walk takes them 28 minutes, how many minutes do they walk after getting their snacks? _____

   Show your reasoning below.

Practice: Missing Quantities—Parts and Wholes

Complete each row with the missing quantities, and explain how you found the answer. The first one is done and explained for you.

	$\frac{1}{4}$	$\frac{3}{4}$	$\frac{4}{4}$
1.	12	12 12 12 12 So $\frac{3}{4}$ is 36	If 12 people is $\frac{1}{4}$, and $\frac{4}{4}$ is 4 × 12, then 48 is the whole, or $\frac{4}{4}$.
2.		12	
3.			12
4.	10		
5.		18	
6.			60

7. When you were given $\frac{3}{4}$ of an amount, how did you find the total or the whole?

Extension: 3/4 of a Million?

The Evening Tribune newspaper is distributed in three towns: Valley Forge, Beantown, and Washington.

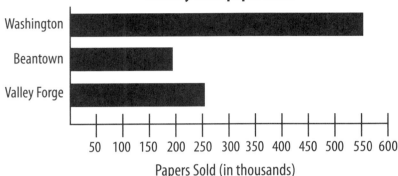

The Evening Tribune
Sunday Newspaper Sales

1. The number of newspapers sold in Valley Forge was

   a. More than $\frac{3}{4}$ of a million.

   b. Less than $\frac{3}{4}$ of a million.

   c. $\frac{3}{4}$ of a million.

   How do you know?

2. The number of newspapers sold in Beantown and Washington together was

   a. More than $\frac{3}{4}$ of a million.

   b. Less than $\frac{3}{4}$ of a million.

   c. $\frac{3}{4}$ of a million.

   How do you know?

3. The number of newspapers sold in Beantown and Valley Forge was

   a. More than $\frac{3}{4}$ of the number sold in Washington.

   b. Less than $\frac{3}{4}$ of the number sold in Washington.

   c. $\frac{3}{4}$ of the number sold in Washington.

   How do you know?

1. A cell-phone plan advertises 750 free minutes. Of those, 200 are daytime minutes. What fraction of the free minutes are on nights and weekends?

   (1) Less than $\frac{1}{4}$

   (2) Exactly $\frac{1}{4}$

   (3) Less than $\frac{3}{4}$

   (4) Exactly $\frac{3}{4}$

   (5) More than $\frac{3}{4}$

2. Alberto is a valet at a mall. He parked 320 cars. He hopes to get tips from at least 3/4 of the drivers. How many drivers would that represent?

   (1) 80

   (2) 160

   (3) 240

   (4) 260

   (5) 300

3. Alice has $12 in her pocket. She wants to spend only 75% of that money on lunch. Looking at the lunch board, what can Alice buy to use exactly 75% of her money?

The Silver Spatula
Lunch Menu

Hot Turkey Sandwich $6.50
American Chop Suey $6.75
Tofu Kabobs $6.00
House Omelette $5.50
Cheeseburger Deluxe $5.50
Hot Dogs and Baked Beans $5.00
Soda .... $1.50 med. / $2.00 lg.
Milk .... $1.50 sm. / $2.50 lg.
Coffee .... $2.25 w/refill

   (1) American chop suey and a large soda

   (2) Hot turkey sandwich and a large milk

   (3) Tofu kabobs and a medium soda

   (4) Cheeseburger Deluxe and a coffee

   (5) House omelette and a large milk

4. Sadie and her husband own the Blossoms Flower Shop. Last week they charted the types of different flowers sold for the week. Which flowers added up to $\frac{3}{4}$ of the total flowers they sold for the week?

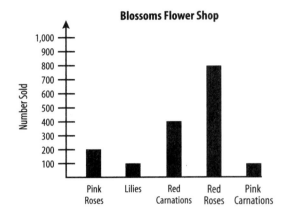

Blossoms Flower Shop

(1) Red carnations

(2) Red carnations and pink roses

(3) Red roses

(4) Red roses and red carnations

(5) Pink roses and lilies

5. Which picture does not show $\frac{3}{4}$?

(1)

(2)

(3)

(4)

(5)

6. Velma lent her son $500 to buy a new car. They agreed he would pay back half the money the first year, a quarter of the money the second year, and the final quarter in the third year. At the start of the third year, how much money will Velma's son owe?

Fraction Stations

> **What benchmark fraction comes to mind?**

You will have a chance in this lesson to solve problems involving fractions that you learned about in the last few lessons. You will travel from one Fraction Station to another, comparing fractions to determine whether they are greater than, less than, or equal to $\frac{1}{4}$, $\frac{1}{2}$, and $\frac{3}{4}$.

Comparing estimates to benchmark fractions helps you describe and make sense of numbers.

Activity: Fraction Stations

Station 1. Data about the Class

1. Write a fraction that describes something about the class data.

2. Locate your fraction on the number line below.

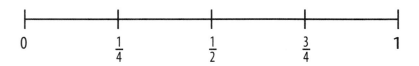

3. Write a sentence with benchmark fractions ($\frac{1}{2}$, $\frac{1}{4}$, and/or $\frac{3}{4}$) to describe the data.

Station 2. Data about Our Neighborhood

1. Write a fraction that describes something about the neighborhood data.

2. Locate your fraction on the number line below.

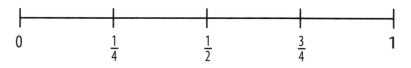

3. Write a sentence with benchmark fractions ($\frac{1}{2}$, $\frac{1}{4}$, and/or $\frac{3}{4}$) to describe the data.

Station 3. A Bookmarked Page

    1. Write a fraction that describes where the bookmark is.

    2. Locate your fraction on the number line below.

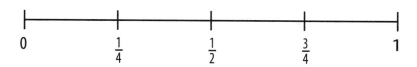

    3. Write a sentence with benchmark fractions ($\frac{1}{2}$, $\frac{1}{4}$, and/or $\frac{3}{4}$) to describe where the bookmark is.

Station 4. Cut the Deck

    1. Cut the deck into two piles.

    2. Look at the piles. Write a sentence using benchmark fractions to estimate the portion of the deck in one of the piles.

    Now count the cards carefully to write an exact fraction to describe that pile.

    4. Place the exact fraction on the number line below.

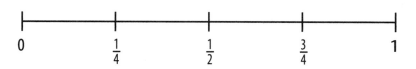

    5. Was your benchmark estimate a good one? Explain why.

Station 5. Pinned Piece of Clothing

1. Use benchmark fractions to describe the approximate location of the safety pin.

2. Now use a ruler or tape measure to get a more exact fraction. Write down the fraction.

3. Place the more exact fraction on the number line below.

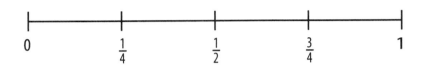

4. Was your benchmark estimate a good one? Explain why.

Station 6. How Full Is the Container?

1. Use benchmark fractions to describe how full the container is.

2. Now measure to determine a more exact fraction.

3. Place the more exact fraction on the number line below.

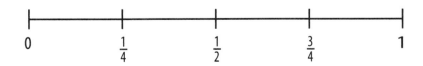

4. Was your benchmark estimate a good one? Explain why.

Practice: Less Than, More Than, Equal to, or Between?

For each graphic below, determine whether the fraction shown is

- Equal to $\frac{1}{4}$, $\frac{1}{2}$, or $\frac{3}{4}$.
- Between $\frac{1}{4}$ and $\frac{1}{2}$ or between $\frac{1}{2}$ and $\frac{3}{4}$.
- More than $\frac{3}{4}$.
- Less than $\frac{1}{4}$.

Write the phrase next to the picture that matches it.

1.

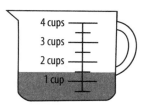

2.

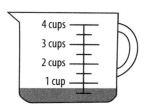

3.

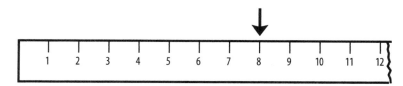

4.

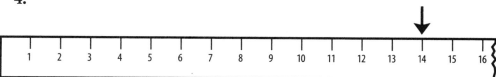

5.

6.

7. Choose three of the pictures from this practice. Explain how you arrived at your answer for each one.

Graphic _____

Graphic _____

Graphic _____

Extension: Describing Data

For each set of data below, compare the fraction of data to the benchmark fractions $\frac{1}{4}$, $\frac{1}{2}$, and $\frac{3}{4}$.

> Digging out information found in graphs, charts, and reports requires careful reading. Make sure you find the right information and that you are working with the right numbers. For instance, is it 1,400 or 14,000 watts used in Problem 1a?

1. Electricity Usage for Electric Appliances

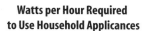

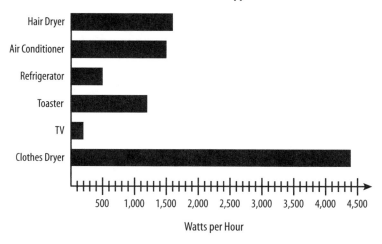

The average home in the United States uses about 28,000 watts of energy each day (24 hours). For Problems a–c, choose one of the following phrases that best describes the fraction of a household's daily total energy use required by the appliance, and write the number on the line.

(1) Less than $\frac{1}{4}$ of the household's total

(2) Between $\frac{1}{4}$ and $\frac{1}{2}$ of the household's total

(3) Between $\frac{1}{2}$ and $\frac{3}{4}$ of the household's total

(4) More than $\frac{3}{4}$ of the household's total

(5) Exactly $\frac{1}{4}$, $\frac{1}{2}$, or $\frac{3}{4}$ of the household's total

a. The watts required by a TV turned on for seven hours a day: _____

b. The watts used by an air conditioner turned on for 16 hours a day: _____

c. The watts used by a clothes dryer for four hours a day: _____

2. Percentages of Donors' Blood Types in the United States

Donors' Blood Types	Percent of Donors with Blood Type
A+ (A, Rh positive)	34%
A− (A, Rh negative)	6%
B+ (B, Rh positive)	9%
B− (B, Rh negative)	2%
AB+ (AB, Rh positive)	3%
AB− (AB, Rh negative)	1%
O+ (O, Rh positive)	38%
O− (O, Rh negative)	7%

Source: American Association of Blood Banks, 2003

For Problems a and b, choose the number of the phrase that best describes the percent.

   (1)   Less than 25% of the total

   (2)   Between 25% and 50% of the total

   (3)   Between 50% and 75% of the total

   (4)   More than 75% of the total

   (5)   Exactly 25%, 50%, or 75% of the total

a. The percent of U.S. blood donors who have type O+ or B+ or AB+ blood: _____

b. The percent of U.S. blood donors who have any negative blood type: _____

3. TV-Watching and Body-Fat Study

Many U.S. children watch a great deal of television and are not active enough, according to medical researchers. A 1998 study by Anderson and others in the *Journal of the American Medical Association* reported on the relationship between the number of hours children watched television daily and their level of physical fitness. The numbers were based on children's own reports.

Overall, 26% of U.S. children watched four or more hours of television per day. Sixty-seven percent watched at least two hours per day. Non-Hispanic Black children had the highest rates of TV watching—four or more hours per day (42%). Boys and girls who watched four or more hours of television each day had greater body fat than those who watched less than two hours per day.

Eighty percent of U.S. children reported being vigorously active three or more times each week. Twenty percent of U.S. children were vigorously active two or fewer times per week. More girls (26%) exercised two or fewer times a week than boys (17%). Vigorous activity levels were lowest among girls, non-Hispanic Blacks, and Mexican Americans.

For Problems a and b, choose the number of the phrase that best describes the percent.

    (1)   Less than 25% of the total

    (2)   Between 25% and 50% of the total

    (3)   Between 50% and 75% of the total

    (4)   More than 75% of the total

    (5)   Exactly 25%, 50%, or 75% of the total

a. The percent of Non-Hispanic Black children watching four or more hours of television each day: _____

b. The percent of boys "vigorously active two or fewer times per week": _____

Extension: Fractions of Billions

The following information appeared in an information bulletin distributed at a health clinic.

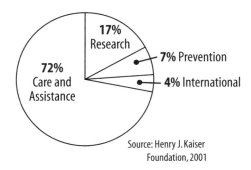

17% Research

7% Prevention

4% International

72% Care and Assistance

Source: Henry J. Kaiser Foundation, 2001

In 2001, the government spent $13.9 billion on Human Immunodeficiency Virus/Acquired Immunodeficiency Syndrome (HIV/AIDS).

1. What fraction of the money went to care and assistance?

2. Which benchmark fraction is closest to that? Explain how you know.

3. What fraction of the money went to research?

4. Which benchmark fraction is closest to that? Explain how you know.

Test Practice

1. Mirabel walked home from the supermarket with her bag of groceries. When she unpacked them, she accidentally dropped a carton of eggs. She discovered that five of the dozen had cracked. What part of the carton of eggs was cracked?

   (1) Less than $\frac{1}{4}$ of the carton

   (2) Between $\frac{1}{4}$ and $\frac{1}{2}$ of the carton

   (3) Between $\frac{1}{2}$ and $\frac{3}{4}$ of the carton

   (4) More than $\frac{3}{4}$ of the carton

   (5) Exactly $\frac{1}{2}$ the carton

2. Jennifer is planning a birthday barbecue for her boyfriend. There will be 18 people at the barbecue. She has 13 ears of corn but wants to make sure that each person gets one. What part of the total ears of corn needed does she already have?

   (1) Less than $\frac{1}{2}$ of what she needs

   (2) Between $\frac{1}{2}$ and $\frac{3}{4}$ of what she needs

   (3) More than $\frac{3}{4}$ of what she needs

   (4) Less than $\frac{1}{4}$ of what she needs

   (5) All she needs

3. Your local ice-cream parlor has a special promotion. Anyone who buys a mint-chip cone gets a free cookie. Thirty-nine of the 50 cones sold in a two-hour shift were mint chip. What part of the total cones sold were mint chip?

   (1) Less than $\frac{1}{4}$ of the cones sold

   (2) Between $\frac{1}{4}$ and $\frac{1}{2}$ of the cones sold

   (3) Between $\frac{1}{2}$ and $\frac{3}{4}$ of the cones sold

   (4) More than $\frac{3}{4}$ of the cones sold

   (5) Exactly 3/4 of the cones sold

4. The corner newsstand sells a wide variety of magazines. Every month they sell 1,100 magazines of which 550 are news magazines. What part of the total magazines sold each month are news magazines?

   (1) Less than $\frac{1}{4}$ of the magazines sold

   (2) Between $\frac{1}{4}$ and $\frac{1}{2}$ of the magazines sold

   (3) Exactly $\frac{1}{2}$ of the magazines sold

   (4) Between $\frac{1}{2}$ and $\frac{3}{4}$ of the magazines sold

   (5) More than $\frac{3}{4}$ of the magazines sold

5. The first rest stop on the way to Cleveland from Cincinnati is 50 miles from Cincinnati. The distance between the two cities is 240 miles. What part of the total distance does the distance from Cincinnati to the first rest stop represent?

   (1) Less than $\frac{1}{4}$ of the total distance between the two cities

   (2) Between $\frac{1}{4}$ and $\frac{1}{2}$ of the total distance between the two cities

   (3) Exactly $\frac{1}{4}$ of the total distance between the two cities

   (4) Between $\frac{1}{2}$ and $\frac{3}{4}$ of the total distance between the two cities

   (5) More than $\frac{3}{4}$ of the total distance between the two cities

6. Hua has a 30-page paper due for her college history class. She has written 18 pages. She has less than _____ (what benchmark fraction) of her paper left to write?

5

One-Tenth

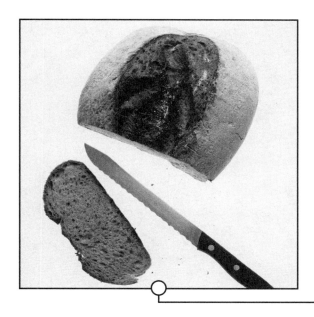

How can you cut a loaf into tenths?

There are 10 one-dollar bills in 10 dollars, 10 sports in a decathlon, and 10 years in a decade. Ten is a friendly number. The same can be said for **one-tenth**.

In this lesson, you will explore the benchmark fraction $\frac{1}{10}$, use what you know to describe it, and find one-tenth of different quantities.

You will also represent one-tenth in several ways, paying attention to when to use which notation.

Activity 1: Show Me $\frac{1}{10}$! Stations

Station 1: Paper Clips

1. What is one-tenth of the whole box?

2. Explain how you know.

Station 2: Your Height

Look around the room to find something that is about a tenth of your height.

1. What object did you select?

    _____.

2. Explain why you think that it is about a tenth of your height.

Station 3: Index Card

- Examine the card that is at the station.

- Decide what one-tenth of that card is.

- Use a pencil to mark the tenth.

1. Explain how you figured out how to mark a tenth of the card.

2. Cut the tenth and tape it below.

3. Is there another way besides the one you used that you could mark the card into tenths? How do you know?

Station 4: One-Quarter

1. What is one-tenth of the quarter at the station?

2. How do you know?

3. How was it different to find one-tenth in this case as compared to finding the tenths at the other stations you visited? Explain.

Station 5: Stamp Collection

The stamps at the station represent one-tenth of someone's stamp collection.

1. How many stamps are in the whole collection?

2. How do you know? Use a drawing, words, or numbers to explain.

Activity 2: Ways to Represent One-Tenth

Write down the different ways to represent one-tenth. Include ones that your classmates share.

Practice: I Will Show You $\frac{1}{10}$!

For each situation below, determine how much one-tenth would equal and show how you know this is true.

You can shade, use number lines or grids, or use numbers to show your thinking.

1.

2. Ten years equal one decade. How many years have passed when 0.1 of a decade goes by?

3. There are 20 pages in a pamphlet. You have read 0.1 of the pamphlet. How many pages have you read?

4.

5. A voting precinct includes 500 voters. If 0.1 of the precinct voters stayed home, how many went out to vote?

Practice: Containers

Mark one-tenth of each of the following containers. Explain how you know you marked one-tenth in each case.

1.

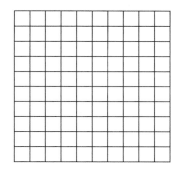

2.

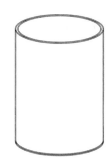

3.

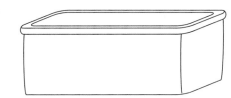

4.

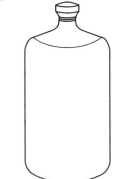

Practice: More or Less?

Decide whether each amount in the following problems is more or less than a tenth. Explain how you know.

1. Zach shows Jamie his three favorite quarters. He says that altogether he has 20 different quarters. Did he show Jamie more or less than one-tenth of his quarters? How do you know?

2. Emily walks her dog each morning. Yesterday she only went around the block four times. Usually she goes around it 10 times. Did she walk more or less than one-tenth of the usual amount yesterday? How do you know?

3. Rachel takes the train into work each morning. She reads a book to pass the time. She is reading a 230-page book. On the train this morning, she read 12 pages. Is that more or less than one-tenth of her book? How do you know?

4. Jon walked half a mile yesterday. Is that more or less than a tenth of a mile? How do you know?

Extension: More Tenths

You have found one-tenth of several different quantities. Now use what you know about finding tenths to find halves and quarters.

1. Ana has $120 saved for a trip she is planning. That, she says, is two-tenths of what she needs to be able to take her trip, hotel and all. How much more must Ana save to have half the money she needs for her trip? How do you know?

2. A coffee shop at the airport sells 240 cups of coffee each hour of a 10-hour day. They keep track four times a day—every quarter—to make sure their sales are on target. How many cups must they sell by the end of the first quarter? How do you know?

1. A marketing specialist noted in 2004 that one in 10 adult Americans did not have a bank account. Which number below could replace the phrase "one in ten"?

    (1) $\frac{1}{10}$

    (2) .01

    (3) $\frac{10}{1}$

    (4) 110

    (5) 1,001

2. Which number does not belong in the following list?

    0.1   10%   .01 $\frac{1}{10}$ .10

    (1) 0.1

    (2) 10%

    (3) .01

    (4) $\frac{1}{10}$

    (5) 0.10

3. A 10-piece pack of gum costs 89¢. Each piece of gum costs about

    (1) 4¢

    (2) 8¢

    (3) 9¢

    (4) 10¢

    (5) 80¢

4. Eighteen eggs are in a large box. One-tenth of the box is about

    (1) one egg.

    (2) two eggs.

    (3) three eggs.

    (4) four eggs.

    (5) nine eggs.

5. Lucy shows Luc two frogs and says they are one-tenth of the frogs in her classroom. How many frogs are in Lucy's classroom?

    (1) 10

    (2) 12

    (3) 15

    (4) 20

    (5) 22

6. Shevan made $14,500 in her ten-month job. What was her monthly income?

More About One-Tenth

Where is the tenth?

In the last lesson, you began thinking about one-tenth, and its various names and representations. In this lesson, you explore visual, fractional, decimal, and percent representations of one-tenth and the connection between one-tenth and nine-tenths. You also examine the relation between $\frac{1}{10}$ and 10%.

Many things in our everyday lives are measured in tenths, such as miles on an odometer, degrees on a thermometer, and decades. It is important to understand tenths and how they relate to other fractions, decimals, and percents. It's also important to pay careful attention to the **decimal point**.

Activity 1: One-Tenth Match

You will receive one or more cards from your teacher. Some of the cards will represent one-tenth; some will not.

1. What quantity does your card represent?

2. Find someone whose card or cards show the same amount (though differently) than yours.

Cards that represent one-tenth:

Activity 2: One-Tenth, Nine-Tenths

You have learned that in mathematics there are many ways to represent the same amount. Try rewriting these common sayings in different ways for the words "nine-tenths" and "one-tenth." Then represent the saying using either a number line or a grid.

> You learned that when you take away $\frac{1}{4}$ of a quantity, you have $\frac{3}{4}$ left because together $\frac{1}{4}$ and $\frac{3}{4}$ make a whole. Together one-tenth and nine-tenths ($\frac{1}{10}$ and $\frac{9}{10}$) also make a whole.

Other Ways to Say "Nine-Tenths"	Other Ways to Say "One-Tenth"
$\frac{9}{10}$	$\frac{1}{10}$
nine out of ten	one out of ten
90%	10%
0.9	0.1
.9	.1

Rewrite each statement two different ways using two different forms of the term(s) "nine-tenths" and/or "one-tenth." One example is done for you.

Use a grid, circle, or number line to illustrate the saying for Part c of the question.

1. **Possession is nine-tenths of the law.**

    a. Example 1: Ownership is 0.9 of the law.

    b. Example 2: $\frac{9}{10}$ of law cases are found in favor of the owner.

    c.

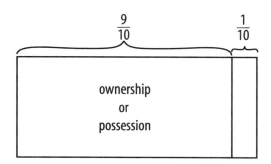

2. **"Nine-tenths of the universe is unseen."**
   (http://www.science-frontiers.com/sf005/sf005p07.htm)

   a.

   b.

   c.

3. **"War is nine-tenths boredom, one-tenth fear."**
   (Evelyn Waugh, author)

   a.

   b.

   c.

4. **"Genius is 10% inspiration, 90% perspiration."**
   (Albert Einstein, physicist)

   a.

   b.

   c.

Activity 3: Why Is 10% Equal to $\frac{1}{10}$?

How would you explain to a friend why 10% equals $\frac{1}{10}$? Use pictures, words, or the grid below to support your explanation.

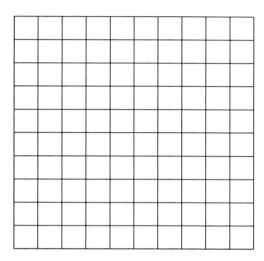

Practice: Make a Gauge

Use either the rectangle or the circle below to make a **gauge** that shows whole numbers from one to five and the tenths in between them. Choose a number that you will mark with an arrow on your gauge. It must be between 0 and 5 and include a decimal, for example, 3.3.

1. Write your number here : _____

2. What might your gauge measure?

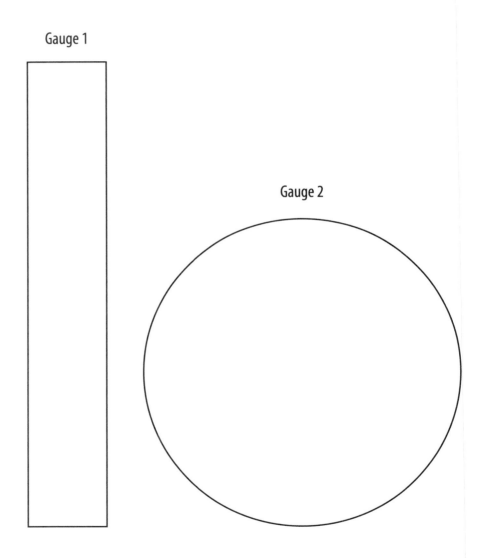

Gauge 1

Gauge 2

Practice: Location, Location, Location

Ava walks her dog, Chi Chi, every day. She makes a three-mile loop around Goose Park.

Use the problem numbers to mark the locations where Ava and Chi Chi stop for each event, and answer the questions.

1. Ava walks 0.1 of a mile, and Chi Chi stops at a hydrant.

2. At 0.5 of a mile, Ava stops and buys a newspaper.

3. Chi Chi loves to stop and look at the lake. They stop after 0.8 of a mile.

4. Another 0.3 of a mile later, they stop in the shade of a grove. How many miles have they walked?

5. After a long stretch of walking, Ava and Chi Chi stop to watch the action on the playing fields. They have walked 2.7 miles.

6. How much farther do Ava and Chi Chi walk before getting back to their starting point?

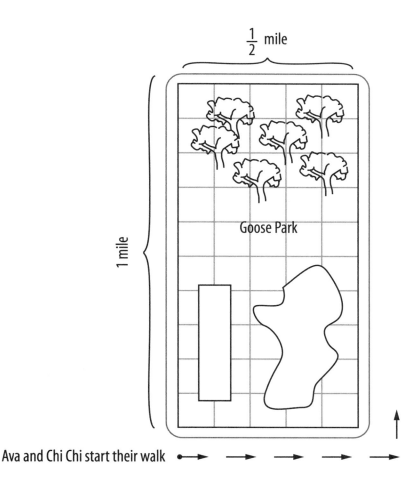

Practice: Gaining Weight

A group of friends made a pact to lose weight. Unfortunately, they gained weight instead!

1. If the original weight of each person listed below increased by 0.1, how much does each one now weigh? Fill in the columns to show what 0.1 of each person's original weight was and what the increased weight is for each of them.

Person	Original Weight in Pounds (lb.)	0.1 of Original Weight	Weight Increased by 0.1	Weight Decreased by 0.1
Claudia	155	15.5 or $15\frac{1}{2}$ lb.	170.5 lb.	139.5 lb.
Salvatore	200			
Jackie	110			
Sue	180			
Calvin	250			
Lamika	135			
Igor	140			

2. If Salvatore, Jackie, Sue, Calvin, Lamika, and Igor each *decreased* his or her weight by 0.1 of the original amount, what would each person weigh now? Fill in the last column with your answers.

Practice: Grid Visions

Part 1

Each block in the grids below represents a person.

Circle one-tenth of each group of people, and complete each equation.

1.

      0.1 of 10 = _____

2.

      0.1 of 20 = _____

3.

Reprinted with permission of *World Education*

0.1 of 100 = _____

Part 2

Complete each equation.

1. 0.1 of 30 = _____

2. 0.1 of 50 = _____

3. 0.1 of 120 = _____

4. What pattern do you notice?

5. How do you find 0.1 of a quantity without a calculator? Give an example.

6. How can you use a calculator to find 0.1 of a quantity? Give an example.

Extension: Reimbursement Changes

There are different ways to say the same thing in math. What are the other names for 0.1 or 0.25? Which of those terms—the decimal, the fraction, or the percent—is most useful to find a part of a given quantity?

In 2004, hourly reimbursement rates for care providers in one state changed. The following table shows some of the jobs and pay rate information for each.

Job Type	For Adult Care			For Childcare		
	2003 Average Pay/Hour	2004 Average Pay/Hour	Change from 2003	2003 Average Pay/Hour	2004 Average Pay/Hour	Change from 2003
Speech Therapist	$108	$243		$108	$173	
Psychologist	$130	$143	+0.1	$131	$206	
Social Worker	$111	$114		$110	$121	+0.1
Nursing Assistant	$ 20	$ 25	+0.25	$ 20	$ 22	+0.1
Occupational Therapist	$ 88	$ 73		$ 88	$116	

Notice the shaded boxes, showing a change in pay from 2003 to 2004. How can you prove that the amount of change is correct in each of the four cases?

1. Psychologist

2. Social Worker

3. Nursing Assistant—Adult Care

4. Nursing Assistant—Childcare

5. Complete the table, filling in all the pay changes from 2003 to 2004.

1. Which of the following does *not* show one-tenth?

   (1)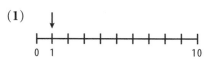

   (2)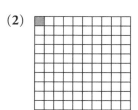

   (3)

   (4)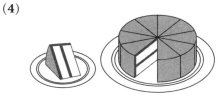

   (5)

2. Which letter location on the number line below most likely represents a 0.1 mark?

   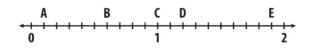

   (1) A
   (2) B
   (3) C
   (4) D
   (5) E

3. One mile measures 5,280 feet. When the odometer (gauge for miles traveled) moves from 10.0 to 10.1, about how many feet *further* have been traveled?

   (1) 5,300
   (2) 530
   (3) 53
   (4) 52
   (5) 5.28

4. Which grid picture represents 0.1?

   A 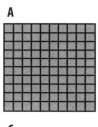   B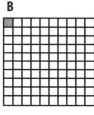

   C 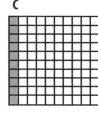   D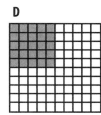

   E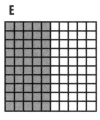

   (1) Grid A
   (2) Grid B
   (3) Grid C
   (4) Grid D
   (5) Grid E

5. Which of the following represents one-tenth?

    (1) 0.01

    (2) 0.1

    (3) 0.11

    (4) 1.1

    (5) 1.01

6. Jean Claude counted all the men and women attending his political rally. There were 260 people altogether, but only 0.1 were men. How many men were at the rally?

Closing the Unit: Benchmarks Revisited

> *How can you use familiar fractions, decimals, or percents to describe this group?*

In this lesson, you will have opportunities to review everything you have learned about benchmark fractions, decimals, and percents. You will solve problems and create a portfolio of your best work. Together with classmates, you will create a benchmark number line.

Are you ready for more fractions, decimals, and percents?

Activity 1: Standing Up

Make notes here as your teacher reads you **statistics** about a group of protest marchers. Write down the fractions, decimals, and whole numbers in the amounts mentioned.

Use the numbers you heard in the statistics and the benchmark fractions to make a diagram, number line, or grid that describes the types of groups participating in the march, such as the number of men, women, and children.

Activity 2: Review Session

Skills Review

Take some time to look through your work.

Try to remember each class you attended.

> What did you do?
>
> What did you learn?

Look back at the *Reflections* and *Vocabulary* sections of your book for more ideas.

- Go back to the practice pages in past lessons.
- Pick a page in each lesson.
- Cover up what you wrote on the page.
- Read the question only.
- Answer out loud or on paper.
- Reread your original answers to refresh your memory and check your work.

Check off the skills you have gained. If you have forgotten how to find an answer, look at your pictures and notes. Reread your notes in *Reflections* for that lesson.

_____ Decide if an amount is more than, less than, or equal to one-half	*Lesson 1*
_____ Write a fraction to show the whole	*Lesson 1*
_____ Find half of a half of a number	*Lesson 2*
_____ Find three-fourths of any number	*Lesson 3*
_____ Find one-tenth of any number	*Lesson 6*
_____ Write one-tenth in other ways	*Lesson 6*
_____ Write one-half using percents and decimals	*Opening*
_____ Write one-fourth using percents and decimals	*Lesson 2*
_____ Write three-fourths using percents and decimals	*Lesson 3*
_____ Put numbers with decimal points in order	*Lesson 5*

Making a Portfolio

Making a portfolio of your best work will help you review the concepts and skills in this unit. Directions for creating a small portfolio follow.

1. Review work you have done in class and on your own. Pick out two assignments you think show your best work or where you learned the most.

2. Make a cover sheet that includes

   - Your name

   - Date

   - Names of the two assignments

   - A picture of a fraction, decimal, or percent of your choice. You can use numbers, number lines, or words along with your picture.

3. For each assignment that you pick,

   - Write a sentence or two describing the piece of work.

   - Write a sentence or two explaining what skills were required to complete the work.

   - Write a sentence or two explaining why you picked this piece of work.

When you have finished, you should have three pieces of paper in your portfolio: one cover sheet and two assignment response pages.

Vocabulary Review

Write a definition or sentence to go with the words you know. Circle the ones you have seen but cannot define, and review your notes to figure out their definitions.

1. One-half

2. One-fourth

3. One-quarter

4. Three-fourths

5. 0.1

6. Decimal point

7. One-tenth

8. 100%

9. 50%

10. 10%

Activity 3: Final Assessment

Complete the tasks on the *Final Assessment*, and then make a Mind Map listing everything you know about benchmark fractions, decimals, and percents. When you finish, compare your first Mind Map, page 2, with your Mind Map from the *Final Assessment*.

What do you notice?

VOCABULARY

Lesson	Terms, Symbols, Concepts	Definitions and Examples
Opening the Unit	one-half, $\frac{1}{2}$	
	fractions	
	decimals	
	percents	
1	benchmark	
	part	
	whole	
2	one-quarter, one-fourth, $\frac{1}{4}$	
3	three-fourths, $\frac{3}{4}$	
4	one-tenth, $\frac{1}{10}$	
5	gauge	
Closing the Unit	statistic	

VOCABULARY *(continued)*

LESSON	TERMS, SYMBOLS, CONCEPTS	DEFINITIONS AND EXAMPLES

REFLECTIONS

OPENING THE UNIT:

I always know if I have half of something because

_____

_____

_____

_____

_____

_____

_____

LESSON 1: More Than, Less Than, or Equal to One-Half

What are some things to remember about comparing an amount to one-half?

_____

_____

_____

_____

_____

_____

_____

LESSON 2: Half of a Half

True things to remember about fourths:

_____

_____

_____

_____

_____

_____

_____

Ways to say and write $\frac{1}{4}$ are

_____
_____
_____
_____
_____
_____
_____

Half of a half means

_____
_____
_____
_____
_____
_____
_____

To find $\frac{1}{4}$ of an amount

_____
_____
_____
_____
_____
_____
_____

LESSON 3: Three Out of Four

Ways to say and write $\frac{3}{4}$ are

_____

_____

_____

_____

_____

_____

To find $\frac{3}{4}$ of an amount

_____

_____

_____

_____

_____

_____

_____

_____

LESSON 4: Fraction Stations

What have you learned since starting the unit *Using Benchmarks: Fractions, Decimals, and Percents*?

_____

_____

_____

_____

_____

_____

_____

I can use this knowledge when

_____
_____
_____
_____
_____
_____
_____

How comfortable are you with fractions at this point?

_____
_____
_____
_____
_____
_____
_____
_____
_____

LESSON 5: One-Tenth

What one-tenth means to me is

_____
_____
_____
_____
_____
_____
_____

I can find one-tenth of any amount by

_____

_____

_____

_____

_____

_____

LESSON 6: More About One-Tenth

What 0.1 means to me is

_____

_____

_____

_____

_____

_____

_____

To find 0.1 of an amount, I can

_____

_____

_____

_____

_____

_____

_____

CLOSING THE UNIT: Benchmarks Revisited

I can use benchmarks (fractions, decimals, and percents) to describe things in the following ways

_____

_____

_____

_____

_____

_____

_____

Final Assessment

What I want to remember about benchmark fractions and decimals ($\frac{1}{2}$, $\frac{1}{4}$, $\frac{3}{4}$, $\frac{1}{10}$, and 0.1) is...

_____

_____

_____

_____

_____

_____

_____